JAMES BRADLEY

GREAT ASTRONOMERS

ROBERT STAWELL BALL

GIFT *Certificate*

TO:

____

____

FROM:

____

____

DATE: ______________________________

JAMES BRADLEY

GREAT ASTRONOMERS

BY

ROBERT STAWELL BALL

D.Sc, LL.D., F.R.S.

(July 1, 1840 – November 25, 1913)

LOWNDEAN PROFESSOR OF ASTRONOMY AND GEOMETRY
IN THE UNIVERSITY OF CAMBRIDGE

DIAMOND BOOKS™

www.diamondbooks.ca

TORONTO, CANADA – 2017

DIAMOND BOOKS - CANADA

Toronto, ON, CANADA
http://www.diamondbooks.ca

BIBLIOGRAPHIC INFORMATION

'JAMES BRADLEY'
Is an extract or a chapter from the book
"GREAT ASTRONOMERS" by Robert Stawell Ball.

FIRST PUBLISHED IN: 1896.

PUBLISHED IN CANADA

Published in Canada by DIAMOND BOOKS - CANADA, an imprint of DIAMOND PUBLISHERS - http://www.diamondpublishers.com

FIRST EDITION: APRIL, 2017.

PAPERBACK EDITION : ISBN: 978-1-988357-62-1

PRINTED IN CANADA

PREFACE
TO THE ORIGINAL EDITION

It has been my object in these pages to present the life of each astronomer in such detail as to enable the reader to realise in some degree the man's character and surroundings; and I have endeavoured to indicate as clearly as circumstances would permit the main features of the discoveries by which he has become known.

There are many types of astronomers—from the stargazer who merely watches the heavens, to the abstract mathematician who merely works at his desk; it has, consequently, been necessary in the case of some lives to adopt a very different treatment from that which seemed suitable for others.

ROBERT STAWELL BALL.

The Observatory, Cambridge.
October, 1895.

NOTE: The full-length of 'Preface' can be found in its original edition of the book "GREAT ASTRONOMERS".

INTRODUCTION

Of all the natural sciences there is not one which offers such sublime objects to the attention of the inquirer as does the science of astronomy. From the earliest ages the study of the stars has exercised the same fascination as it possesses at the present day. Among the most primitive peoples, the movements of the sun, the moon, and the stars commanded attention from their supposed influence on human affairs. The practical utilities of astronomy were also obvious in primeval times. Maxims of extreme antiquity show how the avocations of the husbandman are to be guided by the movements of the heavenly bodies. The positions of the stars indicated the time to plough, and the time to sow. To the mariner who was seeking a way across the trackless ocean, the heavenly bodies .offered the only reliable marks by which his path could be guided. There was, accordingly, a stimulus both from intellectual curiosity and from practical necessity to follow the movements of the stars. Thus began a search for the causes of the ever-varying phenomena which the heavens display.

- ROBERT STAWELL BALL

Many of the earliest discoveries are indeed prehistoric. The great diurnal movement of the heavens, and the annual revolution of the sun, seem to have been known in times far more ancient than those to which any human monuments can be referred. The acuteness of the early observers enabled them to single out the more important of the wanderers which we now call planets. They saw that the star-like objects, Jupiter, Saturn, and Mars, with the more conspicuous Venus, constituted a class of bodies wholly distinct from the fixed stars among which their movements lay, and to which they bear such a superficial resemblance. But the penetration of the early astronomers went even further, for they recognized that Mercury also belongs to the same group, though this particular object is seen so rarely. It would seem that eclipses and other phenomena were observed at Babylon from a very remote period, while the most ancient records of celestial observations that we possess are to be found in the Chinese annals. The study of astronomy, in the sense in which we understand the word, may be said to have commenced under the reign of the Ptolemies at Alexandria. The most famous name in the science of this period is that of Hipparchus, who lived and worked at Rhodes about the year 160 B.C. It was his splendid investigations that first wrought the observed facts into a coherent branch of knowledge. He recognized the primary obligation which lies on the student of the heavens to compile as complete an inventory as possible of the objects which are there to be found. Hipparchus accordingly commenced by undertaking, on a small scale, a task exactly similar to that on which modern astronomers, with all available appliances of meridian circles, and photographic telescopes, are constantly engaged at the present day. He compiled a catalogue of the principal fixed stars, which is of special value to astronomers, as being the earliest work of its kind which has been handed down. He also studied the movements of the sun and the moon, and framed theories to account for the incessant changes which he saw in progress. He found a much more difficult problem in his attempt to interpret satisfactorily the complicated movements of the planets. With the view of constructing a theory which should give some coherent account of the subject, he made many observations of the places of these wandering stars. How great were the advances which Hipparchus accomplished may be appreciated if we reflect that, as a preliminary task to his more purely astronomical labours, he had to invent that branch of mathematical science by which alone the problems he proposed could be solved. It was for this purpose that he devised the indispensable method of calculation which we now know so well as trigonometry. Without the aid rendered by this beautiful art it would have been impossible for any n ally important advance in astronomical calculation to have been effected.

But the discovery which shows, beyond all others, that Hipparchus possessed one of the master-minds of all time was the detection of that remarkable celestial movement known as the precession of the equinoxes. The inquiry which conducted to this discovery involved a most profound investigation, especially when it is remembered that in the days of Hipparchus the means of observation of the heavenly bodies were only of the rudest description, and the available observations of earlier dates were extremely scanty. We can but look with astonishment on the genius of the man who, in spite of such difficulties, was able to detect such a phenomenon as the precession, and to exhibit its actual magnitude. I shall endeavour to explain the nature of this singular celestial movement, for it may be said to offer the first instance in the history of science in which we find that combination of accurate observation with skilful interpretation, of which, in the subsequent development of astronomy, we have so many splendid examples.

The word equinox implies the condition that the night is equal to the day. To a resident on the equator the night is no doubt equal to the day at all times in the year, but to one who lives on any other part of the earth, in either hemisphere, the night and the day are not generally equal. There is, however, one occasion in spring, and another in autumn, on which the day and the night are each twelve hours at all places on the earth. When the night and day are equal in spring, the point which the sun occupies on. the heavens is termed the vernal equinox. There is similarly another point in which the sun is situated at the time of the autumnal equinox. In any investigation of the celestial movements the positions of these two equinoxes on the heavens are of primary importance, and Hipparchus, with the instinct of genius, perceived their significance, and commenced to study them. It will be understood that we can always define the position of a point on the sky with reference to the surrounding stars. No doubt we do not see the stars near the sun when the sun is shining, but they are there nevertheless. The ingenuity of Hipparchus enabled him to determine the positions of each of the two equinoxes relatively to the stars which lie in its immediate vicinity. After examination of the celestial places of these points at different periods, he was led to the conclusion that each equinox was moving relatively to the stars, though that movement was so slow that twenty five thousand years would necessarily elapse before a complete circuit of the heavens was accomplished. Hipparchus traced out this phenomenon, and established it on an impregnable basis, so that all astronomers have ever since recognised the precession of the equinoxes) as one of the fundamental facts of astronomy. Not until nearly two thousand years after Hipparchus had made this splendid discovery was the explanation of its cause given by Newton.

From the days of Hipparchus down to the present hour the science of astronomy has steadily grown. One great observer after another has appeared from time to time, to reveal some new phenomenon with regard to the celestial bodies or their movements, while from time to time one commanding intellect after another has arisen to explain the true import of the facts of observations. The history of astronomy thus becomes inseparable from the history of the great men to whose labours its development is due. In the ensuing chapters we have endeavoured to sketch the lives and the work of the great philosophers, by whose labours the science of astronomy has been created.

We shall commence with Ptolemy, who, after the foundations of the science had been laid by Hipparchus, gave to astronomy the form in which it was taught throughout the Middle Ages. We shall next see the mighty revolution in our conceptions of the universe which are associated with the name of Copernicus. We then pass to those periods illumined by the genius of Galileo and Newton, and afterwards we shall trace the careers of other more recent discoverers, by whose industry and genius the boundaries of human knowledge have been so greatly extended. Our history will be brought down late enough to include some of the illustrious astronomers who laboured in the generation which has just passed away.

- ROBERT STAWELL BALL

TABLE OF CONTENTS

CHAPTERS

**An original Chapter from the book:*
"GREAT ASTRONOMERS", by Robert Stawell Ball.

CHAPTER 9

GREAT ASTRONOMERS

James Bradley

(March 1693 – July 13, 1762)

James Bradley

(March 1693 – July 13, 1762)

CHAPTER 9

GREAT ASTRONOMERS: JAMES BRADLEY

JAMES BRADLEY was descended from an ancient family in the county of Durham. He was born in 1692 or 1693, at Sherbourne, in Gloucestershire, and was educated in the Grammar School at Northleach. From thence he proceeded in due course to Oxford, where he was admitted a commoner at Balliol College, on March 15th, 1711. Much of his time, while an under-graduate, was passed in Essex with his maternal uncle, the Rev. James Pound, who was a well-known man of science and a diligent observer of the stars. It was doubtless by intercourse with his uncle that young Bradley became so expert in the use of astronomical instruments, but the immortal discoveries he subsequently made show him to have been a born astronomer.

The first exhibition of Bradley's practical skill seems to be contained in two observations which he made in 1717 and 1718. They have been published by Halley, whose acuteness had led him to perceive the extraordinary scientific talents of the young astronomer. Another illustration of the sagacity which Bradley manifested, even at the very commencement of his astronomical career, is contained in a remark of Halley's, who says: "Dr. Pound and his nephew, Mr. Bradley, did, myself being present, in the last opposition of the sun and Mars this way demonstrate the extreme minuteness of the sun's parallax, and that it was not more than twelve seconds nor less than nine seconds." To make the significance of this plain, it should be observed that the determination of the sun's parallax is equivalent to the determination of the distance from the earth to the sun.

At the time of which we are now writing, this very important unit of celestial measurement was only very imperfectly known, and the observations of Pound and Bradley may be interpreted to mean that, from their observations, they had come to the conclusion that the distance from the earth to the sun must be more than 94 millions of miles, and less than 125 millions. We now, of course, know that they were not exactly right, for the true distance of the sun is about 93 millions of miles. We cannot, however, but think that it was a very remarkable approach for the veteran astronomer and his brilliant nephew to make towards the determination of a magnitude which did not become accurately known till fifty years later.

Among the earliest parts of astronomical work to which Bradley's attention was directed, were the eclipses of Jupiter's satellites. These phenomena are specially attractive inasmuch as they can be so readily observed, and Bradley found it extremely interesting to calculate the times at which the eclipses should take place, and then to compare his observations with the predicted times. From the success that he met with in this work, and from his other labours, Bradley's reputation as an astronomer increased so greatly that on November the 6th, 1718, he was elected a Fellow of the Royal Society.

Up to this time the astronomical investigations of Bradley had been more those of an amateur than of a professional astronomer, and as it did not at first seem likely that scientific work would lead to any permanent provision, it became necessary for the youthful astronomer to choose a profession. It had been all along intended that he should enter the Church, though for some reason which is not told us, he did not take orders as soon as his age would have entitled him to do so. In 1719, however, the Bishop of Hereford offered Bradley the Vicarage of Bridstow, near Ross, in Monmouthshire, and on July 25th, 1720, he having then taken priest's orders, was duly instituted in his vicarage. In the beginning of the next year, Bradley had some addition to his income from the proceeds of a Welsh living, which, being a sinecure, he was able to hold with his appointment at Bridstow. It appears, however, that his clerical occupations were not very exacting in their demands upon his time, for he was still able to pay long and often-repeated visits to his uncle at Wandsworth, who, being himself a clergyman, seems to have received occasional assistance in his ministerial duties from his astronomical nephew.

The time, however, soon arrived when Bradley was able to make a choice between continuing to exercise his profession as a divine, or devoting himself to a scientific career. The Savilian Professorship of Astronomy in the University of Oxford became vacant by the death of Dr. John Keill. The statutes forbade that the Savilian Professor should also hold a clerical appointment, and Mr. Pound would certainly have been elected to the pro-fessorship had he consented to surrender his preferments in the Church. But Pound was unwilling to sacrifice his clerical position, and though two or three other candidates appeared in the field, yet the talents of Bradley were so conspicuous that he was duly elected, his willingness to resign the clerical profession having been first ascertained.

There can be no doubt that, with such influential friends as Bradley possessed, he would have made great advances had he adhered to his profession as a divine. Bishop Hoadly, indeed, with other marks of favour, had already made the astronomer his chaplain. The engrossing nature of Bradley's interest in astronomy decided him, however, to sacrifice all other prospects in comparison with the opening afforded by the Savilian Professorship. It was not that Bradley found himself devoid of interest in clerical matters, but he felt that the true scope for such abilities as he possessed would be better found in the discharge of the scientific duties of the Oxford chair than in the spiritual charge of a parish. On April the 26th, 1722, Bradley read his inaugural lecture in that new position on which he was destined to confer such lustre.

It must, of course, be remembered that in those early days the art of constructing the astronomical telescope was very imperfectly understood. The only known method for getting over the peculiar difficulties presented in the construction of the refracting telescope, was to have it of the most portentous length. In fact, Bradley made several of his observations with an instrument of two hundred and twelve feet focus. In such a case, no tube could be used, and the object glass was merely fixed at the top of a high pole. Notwithstanding the inconvenience and awkwardness of such an instrument, Bradley by its means succeeded in making many careful measurements. He observed, for example, the transit of Mercury over the sun's disc, on October 9th, 1723; he also measured the dimensions of the planet Venus, while a comet which Halley discovered on October the 9th, 1723, was assiduously observed at Wanstead up to the middle of the ensuing month. The first of Bradley's remarkable contributions to the "Philosophical Transactions" relates to this comet, and the extraordinary amount of work that he went through in connection therewith may be seen from an examination of his book of calculations which is still extant.

The time was now approaching when Bradley was to make the first of those two great discoveries by which his name has acquired a lustre that has placed him in the very foremost rank of astronomical discoverers. As has been often the case in the history of science, the first of these great successes was attained while he was pursuing a research intended for a wholly different purpose. It had long been recognised that as the earth describes a vast orbit, nearly two hundred million miles in diameter, in its annual journey round the sun, the apparent places of the stars should alter, to some extent, in correspondence with the changes in the earth's position. The nearer the star the greater the shift in its apparent place on the heavens, which must arise from the fact that it was seen from different positions in the earth's orbit. It had been pointed out that these apparent changes in the places of the stars, due to the movement of the earth, would provide the means of measuring the distances of the stars. As, however, these distances are enormously great in comparison with the orbit which the earth describes around the sun, the attempt to determine the distances of the stars by the shift in their positions had hitherto proved ineffectual.

Bradley determined to enter on this research once again; he thought that by using instruments of greater power, and by making measurements of increased delicacy, he would be able to perceive and to measure displacements which had proved so small as to elude the skill of the other astronomers who had previously made efforts in the same direction. In order to simplify the investigation as much as possible, Bradley devoted his attention to one particular star, Beta Draconis, which happened to pass near his zenith. The object of choosing a star in this position was to avoid the difficulties which would be introduced by refraction had the star occupied any other place in the heavens than that directly overhead.

We are still able to identify the very spot on which the telescope stood which was used in this memorable research. It was erected at the house then occupied by Molyneux, on the western extremity of Kew Green. The focal length was 24 feet 3 inches, and the eye-glass was 3 feet above the ground floor. The instrument was first set up on November 26th, 1725. If there had been any appreciable disturbance in the place of Beta Draconis in consequence of the movement of the earth around the sun, the star must appear to have the smallest latitude when in conjunction with the sun, and the greatest when in opposition. The star passed the meridian at noon in December, and its position was particularly noticed by Molyneux on the third of that month. Any perceptible displacement by parallax—for so the apparent change in position, due to the earth's motion, is called—would have made the star shift towards the north.

Bradley, however, when observing it on the 17th, was surprised to find that the apparent place of the star, so far from shifting towards the north, as they had perhaps hoped it would, was found to lie a little more to the south than when it was observed before. He took extreme care to be sure that there was no mistake in his observation, and, true astronomer as he was, he scrutinized with the utmost minuteness all the circumstances of the adjustment of his instruments. Still the star went to the south, and it continued so advancing in the same direction until the following March, by which time it had moved no less than twenty seconds south from the place which it occupied when the first observation was made. After a brief pause, in which no apparent movement was perceptible, the star by the middle of April appeared to be returning to the north. Early in June it reached the same distance from the zenith which it had in December. By September the star was as much as thirty-nine seconds more to the north than it had been in March, then it returned towards the south, regaining in December the same situation which it had occupied twelve months before.

This movement of the star being directly opposite to the movements which would have been the consequence of parallax, seemed to show that even if the star had any parallax its effects upon the apparent place were entirely masked by a much larger motion of a totally different description. Various attempts were made to account for the phenomenon, but they were not successful. Bradley accordingly determined to investigate the whole subject in a more thorough manner. One of his objects was to try whether the same movements which he had observed in one star were in any similar degree possessed by other stars. For this purpose he set up a new instrument at Wanstead, and there he commenced a most diligent scrutiny of the apparent places of several stars which passed at different distances from the zenith. He found in the course of this research that other stars exhibited movements of a similar description to those which had already proved so perplexing. For a long time the cause of these apparent movements seemed a mystery. At last, however, the explanation of these remarkable phenomena dawned upon him, and his great discovery was made.

One day when Bradley was out sailing he happened to remark that every time the boat was laid on a different tack the vane at the top of the boat's mast shifted a little, as if there had been a slight change in the direction of the wind. After he had noticed this three or four times he made a remark to the sailors to the effect that it was very strange the wind should always happen to change just at the moment when the boat was going about. The sailors, however, said there had been no change in the wind, but that the alteration in the vane was due to the fact that the boat's course had been altered. In fact, the position of the vane was determined both by the course of the boat and the direction of the wind, and if either of these were altered there would be a corresponding change in the direction of the vane. This meant, of course, that the observer in the boat which was moving along would feel the wind coming from a point different from that in which the wind appeared to be blowing when the boat was at rest, or when it was sailing in some different direction. Bradley's sagacity saw in this observation the clue to the difficulty which had so long troubled him.

It had been discovered before the time of Bradley that the passage of light through space is not an instantaneous phenomenon. Light requires time for its journey. Galileo surmised that the sun may have reached the horizon before we see it there, and it was indeed sufficiently obvious that a physical action, like the transmission of light, could hardly take place without requiring some lapse of time. The speed with which light actually travelled was, however, so rapid that its determination eluded all the means of experimenting which were available in those days. The penetration of Roemer had previously detected irregularities in the observed times of the eclipses of Jupiter's satellites, which were undoubtedly due to the interval which light required for stretching across the interplanetary spaces. Bradley argued that as light can only travel with a certain speed, it may in a measure be regarded like the wind, which he noticed in the boat. If the observer were at rest, that is to say, if the earth were a stationary object, the direction in which the light actually does come would be different from that in which it appears to come when the earth is in motion. It is true that the earth travels but eighteen miles a second, while the velocity with which light is borne along attains to as much as 180,000 miles a second. The velocity of light is thus ten thousand times greater than, the speed of the earth. But even though the wind blew ten thousand times faster than the speed with which the boat was sailing there would still be some change, though no doubt a very small change, in the position of the vane when the boat was in progress from the position which it would have if the boat were at rest. It therefore occurred to this most acute of astronomers that when the telescope was pointed towards a star so as to place it apparently in the centre of the field of view, yet that was not generally the true position of the star. It was not, in fact, the position in which the star would have been observed had the earth been at rest. Provided with this suggestion, he explained the apparent movements of the stars by the principle known as the "aberration of light. "Every circumstance was accounted for as a consequence of the relative movements of the earth and of the light from the star. This beautiful discovery not only established in the most forcible manner the nature of the movement of light; not only did it illustrate the truth of the Copernican theory which asserted that the earth revolved around the sun, but it was also of the utmost importance in the improvement of practical astronomy. Every observer now knows that, generally speaking, the position which the star appears to have is not exactly the position in which the star does actually lie. The observer is, however, able, by the application of the principles which Bradley so clearly laid down, to apply to an observation the correction which is necessary to obtain from it the true place in which the object is actually situated. This memorable achievement at once conferred on Bradley the highest astronomical fame. He tested his discovery in every way, but only to confirm its truth in the most complete manner.

Halley, the Astronomer Royal, died on the 14th January, 1742, and Bradley was immediately pointed out as his successor. He was accordingly appointed Astronomer Royal in February, 1742. On first taking up his abode at Greenwich he was unable to conduct his observations owing to the wretched condition in which he found the instruments. He devoted himself, however, assiduously to their repair, and his first transit observation is recorded on the 25th July, 1742. He worked with such energy that on one day it appears that 255 transit observations were taken by himself alone, and in September, 1747, he had completed the series of observations which established his second great discovery, the nutation of the earth's axis. The way in which he was led to the detection of the nutation is strikingly illustrative of the extreme care with which Bradley conducted his observations. He found that in the course of a twelvemonth, when the star had completed the movement which was due to aberration, it did not return exactly to the same position which it had previously occupied. At first he thought this must be due to some instrumental error, but after closer examination and repeated study of the effect as manifested by many different stars, he came to the conclusion that its origin must be sought in some quite different source. The fact is that a certain change takes place in the apparent position of the stars which is not due to the movement of the star itself, but is rather to be attributed to changes in the points from which the star's positions are measured.

We may explain the matter in this way. As the earth is not a sphere, but has protuberant parts at the equator, the attraction of the moon exercises on those protuberant parts a pulling effect which continually changes the direction of the earth's axis, and consequently the position of the pole must be in a state of incessant fluctuation. The pole to which the earth's axis points on the sky is, therefore, slowly changing. At present it happens to lie near the Pole Star, but it will not always remain there. It describes a circle around the pole of the Ecliptic, requiring about 25,000 years for a complete circuit. In the course of its progress the pole will gradually pass now near one star and now near another, so that many stars will in the lapse of ages discharge the various functions which the present Pole Star does for us. In about 12,000 years, for instance, the pole will have come near the bright star, Vega. This movement of the pole had been known for ages. But what Bradley discovered was that the pole, instead of describing an uniform movement as had been previously supposed, followed a sinuous course now on one side and now on the other of its mean place. This he traced to the fluctuations of the moon's orbit, which undergoes a continuous change in a period of nineteen years. Thus the efficiency with which the moon acts on the protuberant mass of the earth varies, and thus the pole is caused to oscillate.

This subtle discovery, if perhaps in some ways less impressive than Bradley's earlier achievements of the detection of the aberration of light, is regarded by astronomers as testifying even in a higher degree to his astonishing care and skill as an observer, and justly entitles him to a unique place among the astronomers whose discoveries have been effected by consummate practical skill in the use of astronomical instruments.

Of Bradley's private or domestic life there is but little to tell. In 1744, soon after he became Astronomer Royal, he married a daughter of Samuel Peach, of Chalford, in Gloucestershire. There was but one child, a daughter, who became the wife of her cousin, Rev. Samuel Peach, rector of Compton, Beauchamp, in Berkshire.

Bradley's last two years of life were clouded by a melancholy depression of spirits, due to an apprehension that he should survive his rational faculties. It seems, however, that the ill he dreaded never came upon him, for he retained his mental powers to the close. He died on 13th July, 1762, aged seventy, and was buried at Minchinhampton.

COMPLIMENTARY MATERIAL

A BRIEF BIOGRAPHICAL SKETCH

JAMES BRADLEY

- ROBERT STAWELL BALL

James Bradley

(March 1693 – July 13, 1762)

COMPLIMENTARY MATERIAL

A BRIEF BIOGRAPHICAL SKETCH

JAMES BRADLEY

James Bradley FRS (March 1693 – 13 July 1762) was an English astronomer and priest and served as Astronomer Royal from 1742, succeeding Edmond Halley. He is best known for two fundamental discoveries in astronomy, the aberration of light (1725–1728), and the nutation of the Earth's axis (1728–1748). These discoveries were called "the most brilliant and useful of the century" by Jean Baptiste Joseph Delambre, historian of astronomy, mathematical astronomer and director of the Paris Observatory, in his history of astronomy in the 18th century (1821), because "It is to these two discoveries by Bradley that we owe the exactness of modern astronomy. .... This double service assures to their discoverer the most distinguished place (after

Hipparchus and Kepler) above the greatest astronomers of all ages and all countries."

Biography

Bradley was born at Sherborne, near Cheltenham in Gloucestershire, to William Bradley and Jane Pound in March 1693. After attending Westwood's Grammar School[3] at Northleach, he entered Balliol College, Oxford, on 15 March 1711, and took degrees of B.A. and M.A. in 1714 and 1717 respectively. His early observations were made at the rectory of Wanstead in Essex, under the tutelage of James Pound, his uncle and a skilled astronomer. Bradley was elected a fellow of the Royal Society on 6 November 1718.

He took orders on becoming vicar of Bridstow in the following year, and a small sinecure living in Wales was also procured for him by his friend Samuel Molyneux. He resigned his ecclesiastical preferments in 1721, when appointed to the Savilian chair of astronomy at Oxford, while as reader on experimental philosophy from 1729 to 1760, he delivered 79 courses of lectures at the Ashmolean Museum.

In 1742, Bradley was appointed to succeed Edmond Halley as Astronomer Royal; his reputation enabled him to apply successfully for a set of instruments costing GB£1,000; and with an 8-foot quadrant completed for him in 1750 by John Bird, he accumulated at Greenwich in ten years materials of inestimable value for the reform of astronomy. A crown pension of GB£250 a year was conferred upon him in 1752.

Bradley retired in broken health, nine years later, to the Cotswold village of Chalford in Gloucestershire. He had medical attention from a local doctor, and Daniel Lysons from Oxford. He died at Skiveralls House on 13 July 1762. He was buried in the churchyard of Holy Trinity church, Minchinhampton in Gloucestershire.

Works

In 1722 Bradley measured the angular diameter of Venus with a large aerial telescope with an objective focal length of 212 ft (65 m).

Bradley's discovery of the aberration of light was made while attempting to detect stellar parallax.[7] Bradley worked with Samuel Molyneux until Molyneux's death in 1728, trying to measure the parallax of Gamma Draconis.

This stellar parallax ought to have shown up, if it existed at all, as a small annual cyclical motion of the apparent position of the star. However, while Bradley and Molyneux did not find the expected apparent motion due to parallax, they found instead a different and unexplained annual cyclical motion. Shortly after Molyneux's death, Bradley realised that this was caused by what is now known as the aberration of light. The basis on which

Bradley distinguished the annual motion actually observed from the expected motion due to parallax, was that its annual timetable was different.

Calculation showed that if there had been any appreciable motion due to parallax, then the star should have reached its most southerly apparent position in December, and its most northerly apparent position in June. What Bradley found instead was an apparent motion that reached its most southerly point in March, and its most northerly point in September; and that could not be accounted for by parallax: the cause of a motion with the pattern actually seen was at first obscure.

A story has often been told, probably apocryphally, that the solution to the problem eventually occurred to Bradley while he was in a sailing-boat on the River Thames. He noticed that when the boat turned about, a small flag at the top of the mast (a telltale) changed its direction, even though the wind had not changed; the only thing that had changed was the direction and speed of the boat. Bradley worked out the consequences of supposing that the direction and speed of the earth in its orbit, combined with a consistent speed of light from the star, might cause the apparent changes of stellar position that he observed. He found that this fitted the observations well, and also gave an estimate for the speed of light, and showed that the stellar parallax, if any, with extremes in June and December, was far too small to measure at the precision available to Bradley. (The smallness of any parallax, compared with expectations, also showed that the stars must be many times more distant from the Earth than anybody had previously believed.)

This discovery of what became known as the aberration of light was, for all realistic purposes, conclusive evidence for the movement of the Earth, and hence for the correctness of Aristarchus' and Kepler's theories; it was announced to the Royal Society in January 1729 (*Phil. Trans.* xxxv. 637). The theory of the aberration also gave Bradley a means to improve on the accuracy of the previous estimate of the speed of light, which had previously been estimated by the work of Ole Rømer and others.

The earliest observations upon which the discovery of the aberration was founded were made at Molyneux's house on Kew Green, and were continued at the house of Bradley's uncle James Pound in Wanstead, Essex. After publication of his work on the aberration, Bradley continued to observe, to develop and check his second major discovery, the nutation of the Earth's axis, but he did not announce this in print until 14 February 1748 (Phil. Trans. xlv. I), when he had tested its reality by minute observations during an entire revolution (18.6 years) of the moon's nodes.

The publication of Bradley's observations was delayed by disputes about their ownership; but they were finally issued by the Clarendon Press, Oxford, in two folio volumes (1798, 1805). The insight and industry of Friedrich Wilhelm Bessel were, however, needed for the development of their fundamental importance.

ABOUT THE AUTHOR

ROBERT STAWELL BALL

(July 1, 1840 – November 25, 1913)

Sir Robert Stawell Ball was an Irish astronomer who founded the screw theory.

He was the son of naturalist Robert Ball and Amelia Gresley Hellicar. He was born in Dublin.

Ball worked for Lord Rosse from 1865 to 1867. In 1867 he became Professor of Applied Mathematics at the Royal College of Science in Dublin. There he lectured on mechanics and published an elementary account of the science.

In 1874 Ball was appointed Royal Astronomer of Ireland and Andrews Professor of Astronomy in the University of Dublin at Dunsink Observatory.

Ball contributed to the science of kinematics by delineating the screw displacement:

When Ball and the screw theorists speak of screws they no longer mean actual cylindrical objects with helical threads cut into them but the possible motion of any body whatsoever, including that of the screw independently of the nut.

Ball's treatise *The Theory of Screws* (1876) is now in the public domain. His work on screw dynamics earned him in 1879 the

Cunningham Medal of the Royal Irish Academy.

In 1882 Popular Science Monthly carried his article "*A Glimpse through the Corridors of Time*". The following year it carried his two-part article on "*The Boundaries of Astronomy*".

Ball expounded the tides in *Time and Tide*: a Romance of the Moon In 1892 he was appointed Lowndean Professor of Astronomy and Geometry at Cambridge University at the same time becoming director of the Cambridge Observatory. He was a fellow of King's College, Cambridge.

In 1900 Cambridge University Press published *A Treatise on the Theory of Screws*. That year he also published *The Story of the Heavens*. Much in the limelight, he stood as President of the Quaternion Society. He was also President of the Mathematical Association in 1900.

In 1908 he published *A Treatise on Spherical Astronomy*, which is a textbook on astronomy starting from spherical trigonometry and the celestial sphere, considering atmospheric refraction and aberration of light, and introducing basic use of a generalised instrument.

His work *The Story of the Heavens* is mentioned in the "Ithaka" chapter of Ulysses. His lectures, articles and books (e.g. *Starland* and *The Story of the Heavens*) were mostly popular and simple in style.

He died in Cambridge and was buried at the Parish of the Ascension Burial Ground in Cambridge, with his wife Lady Francis Elizabeth Ball. Their children were: Frances Amelia, Robert Steele, William Valentine (later Sir), Mary Agnetta, Charles Rowan Hamilton, and Randall Gresley (later Colonel).

LECTURES

Ball became celebrated for his popular lectures on science. He gave an estimated 2500 lectures between 1875 and 1910 in towns and cities across Britain and Ireland. In 1892, 1898 and 1900 he was invited to deliver the Royal Institution Christmas Lecture. *Astronomy; Astronomy* and *Great Chapters from the Book of Nature.*

FEEDBACK

Now that you have read the book ...

Was it interesting?

Did you enjoy what you wanted to read?
Was there any room for improvement?

Let us know at:

http://www.diamondbooks.ca/feedback

Your feedback is highly appreciated.

Thank you!

NOTES

NOTES

NOTES

NOTES

NOTES

NOTES

NOTES

NOTES

Made in United States
Cleveland, OH
06 November 2026

47647254R00040